Alphabet Homestead

Alphabet Homestead

in the meadow BOOKS

VALERIE A. PARKER

in the meadow BOOKS

ISBN: 978-1-950513-00-0

Written by: Valerie A. Parker
Edited by: Valerie A. Parker
Photography by: Valerie A. Parker
Photography by: Hannah Parker

Thank you God.
I give you all the glory.

This book is dedicated to our Annie — the best little dog in the whole world and the original inspiration behind all the books I've written. We will never forget the love and joy you brought into our lives. You will always have a special place in our hearts, and we will love you always and forever.

And to my dad, who we miss every day— I'm grateful for all the weird outdoorsy knowledge you taught me as a child. It's come in quite handy as a redneck farmer.

Aa

A is for Apples picked
right off the tree.
With this sweet fruit,
so many desserts I can
bake, but apple pie is
my favorite to make!

Bb

B is for Bunny who plays hide-and-seek in the sun. She hops through the grass and has lots and lots of fun!

Cc

C is for Cooking in a homestead kitchen. So many things we can bake—from homemade cookies to chocolate cupcakes!

Dd

D is for Donkey
who guards goats
and sheep. A
hardworking helper
who earns what
he keeps!

Ee

E is for Eggs from hens that forage all day. They lay an egg each morning—but still find time to play!

Ff

F is for Farm Dog who helps with all the chores. She works hard on the farm and never gets bored!

Gg

G is for Goats
that romp in the
pasture all day.
And when they're
tired, they munch
on hay!

Hh

H is for Hens, the farmer's helpers. They lay eggs and gobble up bugs. Uh-oh! One's in the garden beds! Look where she dug!

Ii

I is for Insects like spiders and ladybugs too. Grasshoppers hop along helping farmers all the day through!

Jj

J is for Jam from fruit picked in summer's sun. It's made in our kitchen and enjoyed all year long!

Kk

K is for Kitten who finds a new friend in the barn. They both love to nap and help out on the farm!

Ll

L is for Local berries from a farm down the road. We freeze them for winter and eat them when it snows!

Mm

M is for Moo, a cow in the pasture all day. She munches grass and makes milk—what do cows say?

Nn

N is for Nest where baby bunnies snuggle together. With mama nearby, they're safe in all kinds of weather!

Oo
O is for Organic Gardening. With my two hands, tiny seeds I sow. Right into the dirt, then watch the plants grow!

P is for Pigs who helps on the farm.
They root in the dirt with their
snouts, not their arms!

Q is for Quack, the sound ducks make. They like the grass but prefer water to splash!

Rr

R is for Rooster, the farmer's morning alarm. He starts the day crowing, brave and full of charm. He protects the hens and follows them along!

Ss

S is for Sheep, known for wool and cheese. They munch on grass and graze with ease. When tired, they rest and chew their cud, and then bleat a song for those they love!

Tt

T is for Tree, a
home for wildlife.
They give us shelter,
food and air, and
help all living things
everywhere.

Uu

U is for Utility Vehicle, a farmer's helper truck. It carries dirt, tools, and hay through the fields and muck!

Vv

V is for Vegetable Soup. With vegetables and herbs like onions, mushrooms, and sage, it keeps us warm and healthy on cold winter days.

Ww

W is for Wildflowers growing tall in the meadow. Autumn winds carry their seeds as they go. Squirrels and mice find them scattered below.

X is on an old barn door. We go in and out to care for horses, and there's always lots of chores!

Yy

Y is for Yarrow growing around the farm. A perennial plant with healing charm. It comes back each year and helps insects too. It blooms in all colors, even blue!

Zz

Z is for Zinnia. In cold climates, it's an annual planted each year. With many colors, varieties, and sizes, their beauty is clear!

Also by Valerie Parker

The Photographers

Valerie Parker and her daughter Hannah jointly managed Sweet Meadows Farm together for many years (now Sowing Seeds Nursery & Flower Farm). It is a small, diversified permaculture farm and self serve farm stand in South Burlington, Vermont. She and her daughter have had quite the menagerie of animals over the years, including chickens, a very spoiled crew of ADGA Registered Nigerian Dwarf goats, Holland Lop bunnies, various other critters and a program to rescue numerous cats from Vermont, the south and abroad. Valerie holistically raised and midwifed the delivery of over 75 goat kids, including many breech births and even sets of quintuplets over twelve years. Over the years, she and Hannah documented and took countless photos of the farm animals, babies, and surrounding wildlife, nature and beautiful Vermont landscapes. Valerie coached many families over the years on how to raise goats and farm animals holistically and start up homesteading and permaculture with animals. In addition to the day to day of running a nursery and farm stand, she is a published writer and has her own independent publishing company *In the Meadow Books.* Valerie has studied holistic medicine and regenerative homesteading for over twenty years, designs websites and dabbles in various other creative work. She is very involved in her local church, and supports various charities.

Hannah had a camera placed in her hands starting at age nine and has never looked back! Self taught, she has always had a passion for animals, especially rescues. In 2023, Hannah moved over to human photography and you can find her cinematic storytelling style photography on Instagram @hannahparkerphotography. She specializes in couples shoots, elopements and portraits. People travel from all over to do shoots with her and she stays busy with creative projects year round in Vermont and surrounding states.

www.ingramcontent.com/pod-product-compliance
Lightning Source LLC
LaVergne TN
LVHW070626061126
851857LV00002B/21

* 9 7 8 1 9 5 0 5 1 3 0 0 0 *